TRAVELER'S GUIDE TO THE
Solar System

Patricia Barnes-Svarney

Sterling Publishing Co., Inc.   New York

To my parents and my husband, Tom, who helped me pursue my dream of space,
and to Nancy Mack, who gave me the friendly push to start

Library of Congress Cataloging-in-Publication Data

Barnes-Svarney, Patricia.
    Traveler's guide to the solar system / by Patricia Barnes-Svarney.
      p.   cm.
    Includes index.
    Summary: Takes the reader on a tour of the solar system, describing asteroids
and each of the planets.
    ISBN 0-8069-8672-7
    1. Solar system—Juvenile literature.   [1. Solar system.]
  I. Title.
QB501.3.B37   1993
523.2—dc20
                                      93-17313
                                      CIP
                                      AC

2 4 6 8 10 9 7 5 3 1

First paperback edition published in 1994 by
Sterling Publishing Company, Inc.
387 Park Avenue South, New York, N.Y. 10016
© 1993 by Patricia Barnes-Svarney
Distributed in Canada by Sterling Publishing
℅ Canadian Manda Group, P.O. Box 920, Station U
Toronto, Ontario, Canada M8Z 5P9
Distributed in Great Britain and Europe by Cassell PLC
Villiers House, 41/47 Strand, London WC2N 5JE, England
Distributed in Australia by Capricorn Link (Australia) Pty Ltd.
P.O. Box 6651, Baulkham Hills, Business Centre, NSW 2153, Australia
Printed and bound in Hong Kong
All rights reserved

Sterling ISBN 0-8069-8672-7 Trade
0-8069-8675-1 Paper

The author wishes to thank the National Aeronautics and Space Administration
(NASA) and the Jet Propulsion Lab of NASA for supplying most of the photographs
for this book, and to acknowledge also the courtesy of Dr. Peter Thomas and the
cooperation of the European Space Agency and the Space Telescope Science
Institute for the use of their photos.

CONTENTS

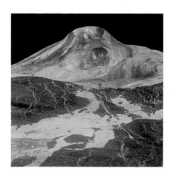 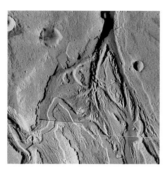 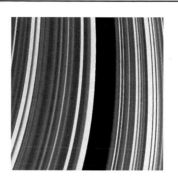 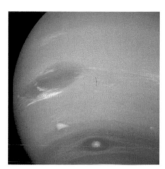

WHERE ARE WE GOING?

Some are dusty and dry; others are cold and icy. But they all have features that make them unique. They are the nine planets and 63 moons (and maybe more) of our solar system!

Where have we traveled in the solar system? Twelve astronauts have walked on the moon. The Viking spacecraft landed on Mars and sent pictures back to Earth. The Venera spacecraft took the first photographs of the rocks on Venus. Other spacecraft, such as the Mariners, Pioneers, and Voyagers, have sent back photos of other planet and satellite surfaces. In fact, Pluto and its moon, Charon, are the only planetary bodies we have not yet visited.

By looking at the data from these planetary probes, we know a great deal about our solar system. By looking at the pictures and stretching our imagination, we can understand what it would be like to walk on the plains of Mars, plunge through the clouds of Saturn—or even watch the wild volcanoes on Jupiter's moon Io.

How do we use our imagination to "go" into space? We have all been on a beach before and know how it feels when the sand blows around our feet. We can feel the small grains strike our legs and cover our toes. No doubt, a small dust storm on Mars would "feel" the same way—except we would be wearing space suits! What about the clouds rushing around Saturn? We know what the clouds of a storm look like as we watch a fast-moving thunderstorm. This gives us a good idea of how some clouds may move on Saturn.

We may not know everything about the members of the solar system. But, based on what we know about our own planet, we can guess—and use our imagination.

So let's take a space hike. We will base our views of the planets and satellites on pictures and data from spacecraft and Earth telescopes. Let's visit some of the best—and some of the worst—places in our solar system.

In this photo, we can see how small our planet is in comparision to the gas giant planets—and even Saturn's rings.

MERCURY IS BURNING IN THE SUN!

Look around you—it looks as if the Sun is everywhere!

Notice how it covers almost one quarter of the sky—over three times larger than the Sun as seen from Earth. We are on Mercury, the second smallest planet in our solar system. It is the closest one to the Sun—and one of the solar system's worst places to visit.

What would it be like to live on Mercury? There is no air to scatter the light and make a bright blue sky as on Earth. There is no wind or water. The heat from the Sun bakes everything on one side of the planet, and everything freezes on the other, dark side. No plants, animals, or humans could live on the small planet without special protection from the heat, cold, and lack of air.

WHAT MAKES UP MERCURY?

What does Mercury look like inside? Scientists believe that the planet consists of only a few layers, like a seed in a pod. The inner "seed," called the core, is probably made up of the metals iron and nickel while the outer "seed," called the crust, is made up of rocks probably like those found on Earth.

Mercury may have had an atmosphere once, when the solar system was first formed. But because Mercury is so small and has less gravity than Earth, the intense heat from the Sun probably boiled away the planet's original blanket of air long ago.

The small planet Mercury is covered with thousands of craters. They range in size from a few feet to hundreds of miles in diameter.

Fast Times around the Sun

Years on the small planet would go quickly compared to Earth years, as Mercury orbits the Sun in 88 Earth days. If you were 10 years old on Earth, you would be just over 40 Mercury-years old. In other words, on Mercury you would have traveled forty times around the Sun.

Our days on Mercury would drag. The planet's day is equal to around 59 Earth-days—an Earth-month of light and an Earth-month of darkness. Just think how much you could do before dark—but could you really sleep that long?

Mercury is around 3,030 miles (4,880 kilometres) in diameter—half again as large as our own Moon. Mercury's gravity does not hug us to the surface as it does on Earth. A space traveler who weighs 100 pounds on Earth would weigh 39 pounds on Mercury. Throwing a softball on Mercury would be so easy you could send the ball into orbit. A jump ball in a game of basketball would allow you to leap 10 feet (3 metres) in a single bound!

Let's look closer: Why are there so many deep wrinkles and cracks in the surface? Maybe Mer-

One of the largest craters on Mercury is the Caloris Basin, measuring over 800 miles (1,300 kilometres) in diameter—about the distance between Chicago and New York.

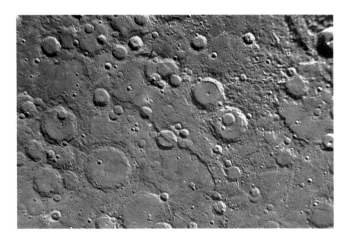

Some of the canyon walls on Mercury are over 300 miles (500 kilometres) long and 2 miles (3.2 kilometres) high.

cury used to be a little bigger billions of years ago when it first formed and it has since shrunk, the way a dried apple shrinks and cracks in the Sun.

Hottest or Coldest?

Mercury holds the record in the solar system for having the largest temperature difference between the day and night sides of the planet. On the dark side, the temperature can drop to about −274 degrees Fahrenheit (−170 degrees Celsius), or 7 times colder than our refrigerator freezers. On the sunlit side, the temperature can soar to 662 degrees Fahrenheit (350 degrees Celsius) or higher. Mercury is definitely too hot and too cold to enjoy any summer or winter sports or activities.

It would be hard to see the stars from the sunlit side of Mercury because the Sun covers so much of the planet's sky. But Mercury's nighttime sky is filled with stars and planets. The lack of a thick layer of air, or atmosphere, makes it even easier to see objects in our universe.

Would we see any life on Mercury? Probably not. Life as we know it needs air, in order to breathe, and moderate temperatures. If there is life on Mercury, it would be very different from what we are used to on the Earth!

MERCURY FROM EARTH

Mercury is one of the brightest objects in our sky. Yet, because it is so close to the bright Sun when it is observed from Earth, many people have never seen it.

The small planet has phases, just like our Moon, and is only visible from Earth about six or seven times a year. It can appear as an "evening star" after dusk or as a "morning star" before sunrise. In fact, the ancients did not realize that the morning star was the same planet as the evening star. The Greeks called the planet "Mercury" when they saw it in the evening and "Apollo" when they saw it in the morning.

Mercury occasionally passes between the Earth and the Sun. This is called a transit, where the small planet, measuring only one percent of the diameter of the Sun, slowly crosses the Sun's face.

WHAT'S UNDER THE CLOUDS OF VENUS?

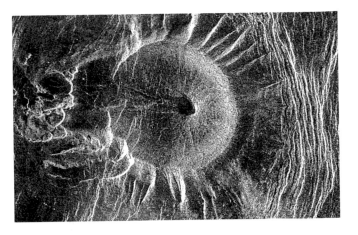

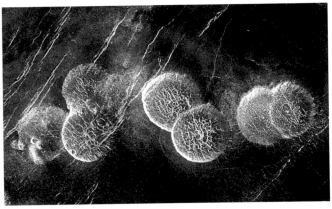

Venus, the second planet from the Sun, is called our "sister" planet. But if we landed on Venus, we would notice that it does not have much in common with Earth except for its size. It is constantly covered with thick, yellow clouds and the surface is extremely hot and dusty—definitely on the list as one of the worst places in the solar system to visit.

Hot Enough to Melt Lead

Venus has the highest surface temperatures of any planet in the solar system. Measured at 900 degrees Fahrenheit (482 degrees Celsius), it is well above the melting point of lead.

Why is Venus so hot? If we had seen Venus and Earth a long time ago, we would have noticed that the atmospheres of both planets were filled with carbon dioxide from exploding volcanoes. On

(Top) **This volcano, nicknamed the Tick, is where hot lava once spewed from deep below the surface. The dome-like hills *(below)* are about 15 miles (24 kilometres) across, and may have formed when sticky lava oozed into a pancake-like puddle.**

(Opposite) **If you strip away Venus' clouds, you will see evidence of craters, channels, and volcanoes, as seen in this picture from the Magellan spacecraft.**

11

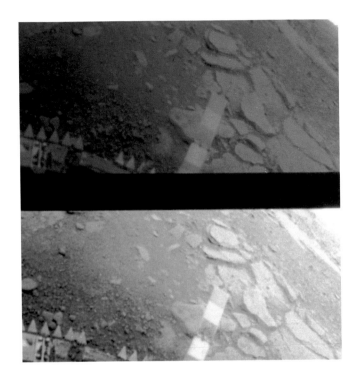

Now try walking. What a challenge! Not only will the thick atmosphere be difficult to walk through, but the slightest breeze will push you back.

Look at the creamy yellow clouds. They are so bright from the Sun's reflection that we have to shade our eyes. If we look closely, the highest clouds seem to be raining. These clouds are made of sulfuric acid and they release burning "acid-drops." But we don't need an umbrella—because of the high heat, the acid-drops evaporate before they ever reach the ground.

(Top) **This is what Venus would look like if you stood on the surface;** *(bottom)* **the surface of Venus as seen in a bright, white spotlight.**

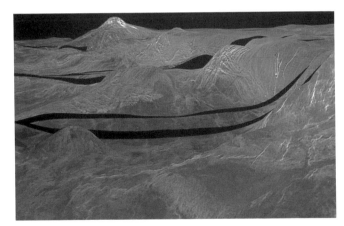

Danu Montes rises around ½ mile to 2 miles (1 to 3 kilometres) above the plains. It is believed to have formed when the area was pushed up by great forces (the black lines are where there was no data collected).

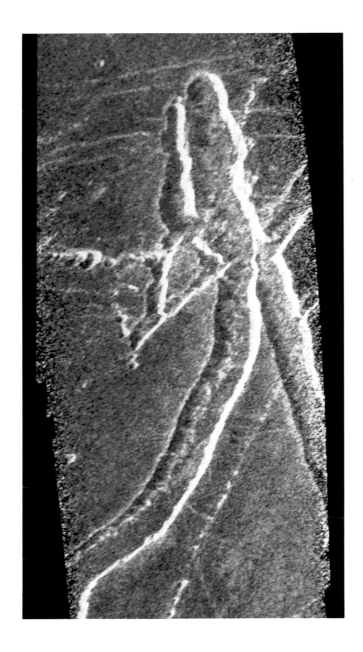

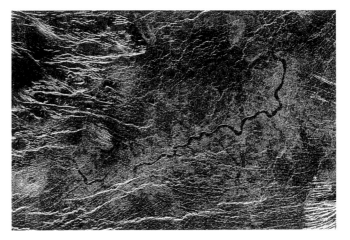

The channels on Venus look almost like river channels on Earth, but, instead of water, they were probably formed by hot lava.

(Left) This trough on Venus is often called the "Gumby feature." The half-mile (about 1-kilometre)-deep trough is thought to be a channel that once carried hot lava.

Whistles and Hums

Have you ever listened to the hums and whistles on a short-wave radio? That is what it sounds like on Venus. Every day, thousands of these whistles can be heard. No one knows—but these "signals" may be from streaks of lightning or strange waves that run through the upper atmosphere.

Look into the sky. If we could see between the thick yellowish clouds, we would see the stars in pretty much the same positions as from the Earth. But rather than marching east to west, the constellations—and the Sun—would move from west to east. This is because Venus rotates in a different direction than Earth and almost all the other planets.

If you did manage to survive on Venus for a while, you would be in for a surprise: Venus races around the Sun in only 225 Earth days. So if you were 10 Earth-years-old, you would be around 16

Maat Mons is a 5-mile (8-kilometre)-high volcano on Venus. It is surrounded by cracked and fractured plains.

VENUS FROM EARTH

At certain times, Venus is brighter than any planet or star in the Earth's sky (except for the Sun, of course). It is so bright that it can often be seen during the day and can even cast a shadow at night.

Because the planet is inside the Earth's orbit, Venus has phases, much like our Moon. When Venus is the brightest to an observer on Earth, it is in the crescent phase; and when it is in full phase, it does not look very bright. (This is just the opposite of our Moon, where its full face is the brightest in our sky and the crescent is not as bright.) The crescent of Venus is brighter because the planet is closer to the Earth at that phase.

NEW VIEWS OF VENUS

What happens when you send a spacecraft with radar to a planet covered with clouds? You get "pictures" that are almost as good as if the clouds were not there!

Long ago, people believed that Venus, because it was covered with clouds, had great oceans. When the Pioneer Venus Orbiter spacecraft arrived at Venus in 1978, it sent back radar pictures of great mountains. In the early 1990s, the Magellan craft returned even better radar pictures of long canyons, strange broken sections of crust, and huge volcanoes and craters. These pictures show that the planet has had volcanic eruptions, many impacts from space objects, and movements of its crust—but no oceans.

Venus-years-old. But your days on Venus would go much more slowly. One day on Venus is equal to 243 Earth days (18 days longer than Venus' year). It would be nice to have a whole Venus day to finish your homework!

Would there be life on Venus? It is hard to say: it seems too hot for any type of life we know. The pressures would also be too great to support Earth animals, plants, and humans. Could a kind of life evolve that would not be affected by such harsh conditions?

There are plenty of craters on Venus that were formed by the impact of large space objects, including Howe Crater. The bright material around the craters is where the surface rock splashed up, then settled on the surface.

Chapter Three

ON TO MARS

How large is Valles Marineris? It would stretch all the way across the United States!

(Opposite) A close look at the red planet Mars showing the vast Valles Marineris canyon.

Let's move on to Mars, where the days, at 24 hours and 37 minutes, are just a little longer than on Earth. Picture a planet where the sky is pink all day and deep blue at sunrise and sunset. The pink skies are caused by the reflection of this red planet's iron-rich dust.

Mars is smaller than the Earth, about half the size, but it has many similarities to our planet. You probably have figured out that because Mars has a pink sky, it must have an atmosphere. You are right! It is made of almost pure carbon dioxide— but is much thinner than the Earth's blanket of gas. If we were to walk on the planet, we would not be able to breathe the thin and poisonous air.

Mars also has four seasons like the Earth. Because the planet orbits the Sun in 687 days, each season on Mars lasts close to 6 Earth-months. There are no fall leaves or spring growth as on Earth, and no rain, sleet, or snow. Seasons on Mars are marked by frost on rocks, the strength of the dust storms, and the changing of the polar caps (the ice caps get larger in the Martian winter and smaller in the summer).

Big Features on a Small Planet

Look around us: we can see tall volcanoes many times larger than any on Earth. One volcano, rising 79,000 feet (24,000 metres) above the surrounding Martian floor, is called Olympus Mons, and it is believed to be the largest volcano in the solar system. It is close to three times taller than our Mount Everest, which stands at 29,028 feet (8,850 metres)!

Blocks of red and black volcanic rocks pocked with holes are strewn across Martian fields. Notice the canyons on Mars. Many of them are deeper than our Grand Canyon, including Valles Marineris, a canyon that would stretch across the entire United States or the Australian continent!

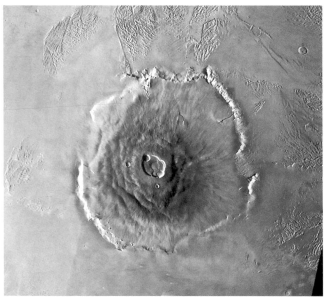

(Left top) The largest feature on Mars is Olympus Mons, a volcano that is larger than any on Earth, and may be the largest in the solar system!

(Bottom) A bird's-eye view of Olympus Mons, complete with small craters that formed after the eruption of the volcano.

Did you notice the two white patches on either end of this red planet? Like Earth, Mars has two polar caps. The south pole is small, only 497 miles (800 kilometres) in diameter, and made of frozen carbon dioxide (what we call "dry ice"). The north pole is made up of mostly water ice and is shaped in a striking spiral pattern.

Each pole grows and shrinks according to the Martian seasons, depending on whether the Sun is closer or farther away. In the Martian summer, when the planet comes closer to the warming Sun, the polar caps shrink as the ice evaporates. This change in temperature also causes high winds that push red dust particles into the air, creating blinding dust storms.

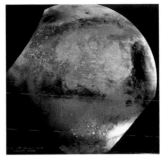 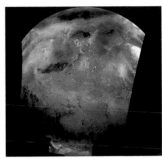

(Above left) **This side of Mars, called the Schiaparelli Hemisphere, shows heavily cratered terrain and the Hellas impact basin at the lower right with "dry ice" frost.**

(Above right) **On the opposite side of Mars is the Cerverus Hemisphere, filled with craters and some volcanoes.**

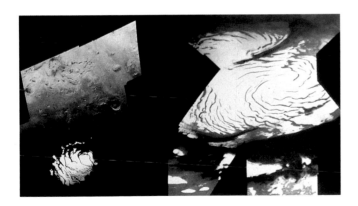

(Left) **There are ice caps on Mars: the smaller ice cap found in the south during the summer is made of carbon dioxide, or dry ice; the larger northern cap seems to be made of mostly water ice.**

A Nice Day on Mars?

The winds often reach more than several hundreds of miles per hour, stirring up the dust. Long tornado-like towers of wind and Martian soil—called "dust devils"—travel at speeds of up to 100 miles (161 kilometres) per hour. Have you ever seen a dust devil on Earth? Next time you are in a parking lot on a hot, dry day, notice how the wind picks up dust, leaves, and dirt, and swirls them around. You are watching a miniature dust devil!

On Mars, the temperatures are much colder than on Earth, so let's hope our spacesuit has a heater! The temperatures on Mars vary from day to night and summer to winter. At the equator, where it is warmest on Mars, temperatures can reach 81 degrees Fahrenheit (27 degrees Celsius) during the Martian summer and can dip to −148 degrees Fahrenheit (−100 degrees Celsius) in the winter. The polar temperatures may drop to −220 degrees Fahrenheit (−140 degrees Celsius).

In spite of the harsh Martian environment, it also may be the one place in the solar system where we will one day build a space colony. What would we expect if we lived in a Martian city? We would need special spacesuits to protect us from the cold and lack of breathable air. We would need shelter from the strong winds and dust storms.

We would also have to have lead weights in our boots to weigh us down, for traction and to walk comfortably on Mars. Why? Because the gravity is very low on Mars. In fact, if you weighed 100 pounds on Earth, you would only weigh 39 pounds on Mars!

Argyre Basin was made by the impact of a very large space object, creating a depression 435 miles (700 kilometres) across.

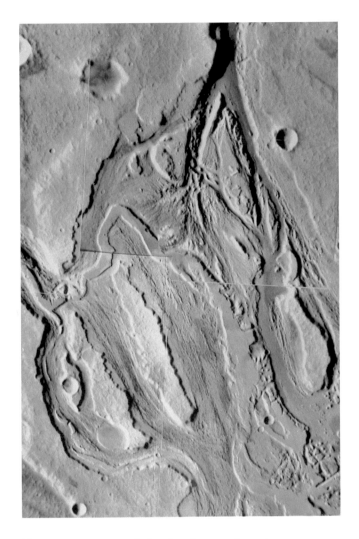

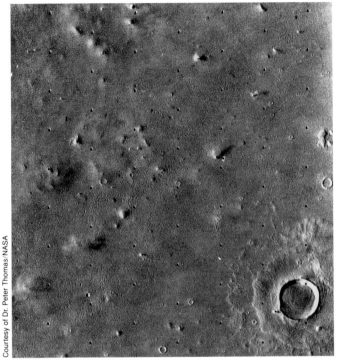

Courtesy of Dr. Peter Thomas/NASA

Some areas have dust devils, tall columns of swirling dust, that occur in the Martian summer. Notice the tall shadows of the dust devils.

Between some of the basins and craters on Mars are channels much like river channels on Earth. Where did the water go? Scientists still don't know!

DO YOU BELIEVE IN MARTIANS?

Have you ever seen Mars through a telescope? You may see features that resemble strange shapes and patterns. If you looked carefully, you would realize that these shapes come from the way light reflects off craters, deep canyons, the long curved channels, and other features on the Martian surface.

In the early 1900s, many astronomers believed they saw linear markings and thought the lines were evidence of living beings on Mars. Many science fiction stories have been written about Mars and the Martians. We know now that there are no Martians on Mars—at least none that could pass for human. If there is some type of life there, it may be too small to be detected by our spacecraft.

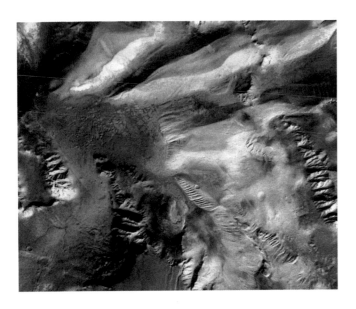

Candor Chasma shows that there are many strange features on Mars, including steep walls and areas that look like landslides.

(Left) **The Viking spacecraft took these two pictures of Mars, complete with volcanic rocks and dust. The bottom photo, taken one morning during a Martian winter, shows white frost covering the rocks.**

Two Tiny Moons

Let's spend a quiet evening on Mars. As we look towards the night sky, we see a blue-white crescent close to the horizon. It is the Earth, but it seems no larger than Jupiter does from Earth. Suddenly, a potato-shaped moon appears on the horizon. It is Phobos. This moon travels around the planet in 7 hours and 39 minutes. A second moon, Deimos, has the same elongated shape and moves slowly around Mars in 30 hours and 18 minutes—6 hours longer than the Martian day! Not only does Phobos pass Deimos in the Martian nighttime sky, it also is going faster than the planet's rotation, so it appears to go in the opposite direction of the other planets and stars.

Phobos and Deimos are not typical of the moons in the solar system. They are too small and too irregular in shape for moons. They are more like asteroids, rocky bodies that also look like potatoes in space. Perhaps the gravity of Mars pulled and captured the two moons from the nearby Asteroid Belt.

As we look at Phobos, we see that it is covered with tiny grooves, small craters, and dust. It is also a moon in trouble. Because of Phobos' low orbit, at only 3,718 miles (5,982 kilometres) above Mars, the planet's gravity is slowly pulling Phobos

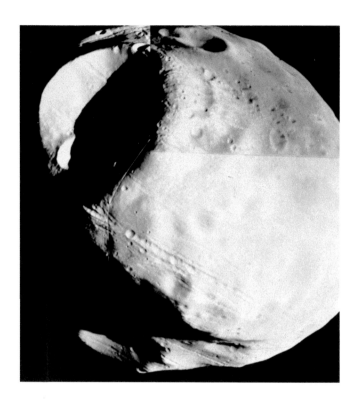

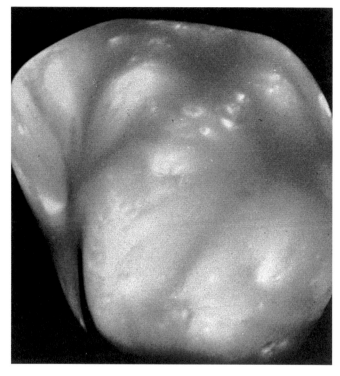

(Above) **Phobos is the largest of the Martian moons. Notice the large crater Stickney—a 6-mile (10-kilometre)-wide hole on the small moon!**

(Above right) **Deimos is the smaller moon of Mars. This photo shows Deimos at "full moon" phase, with a shape that has been referred to as a potato or marshmallow!**

towards it. Some scientists believe that Phobos will one day crash into Mars creating a 62-mile (100-kilometre) crater; others think that the moon will break up into a ring of debris around Mars. But don't stand there and wait—neither will happen for maybe one hundred million years.

Deimos also is covered with grooves, small craters, and dust. It is around 7 miles (11 kilometres) in diameter. Deimos is so small that if we were to run 7 miles (11 kilometres) per hour on the tiny moon, we would launch into space!

Phobos also has an amazing feature on its surface—what looks like a bull's eye. It is really a crater named Stickney. The crater is so large, it covers close to one third of the moon.

Mars Needs Air!

Will we find life on the red planet or its moons? Spacecraft sent to fly by the moons showed them to be rocky bodies devoid of life. Mars does not appear to have life either, but the search results from the Viking spacecraft landers were inconclusive.

Next time, when we go to the red planet, it may be a good idea to land at the poles. After all, life, even if it is microscopic, like bacteria, may be able to survive in the water there. We may even try digging into the soil below the many channels. Who knows—we may find fossils, evidence of ancient life, deep under the Martian surface!

WATER, WATER... BUT WHERE?

On Earth, river channels carry tons of water each year, eventually emptying into the oceans. If you look at a river, it takes a certain path, flowing over the land, cutting its way into the earth. There also are long winding channels on Mars, some larger than the Earth's Amazon basin in South America.

But where is the Martian water? There is a small amount of water on the red planet, trapped in the permanently frozen ground, in the ice caps, in thin wispy clouds, and in frost on the ground on a Martian morning. But there is no liquid water that would explain the huge channels that crisscross the planet.

Scientists do not know how these channels developed. Did the frozen water in the crust somehow melt, sending tons of water to scour the land? How did the water unfreeze? Did one of the volcanoes on the planet become active, heating up the area? Did an asteroid or comet strike Mars and heat up the surface? Or was the Martian climate warmer at one time? No one knows for sure.

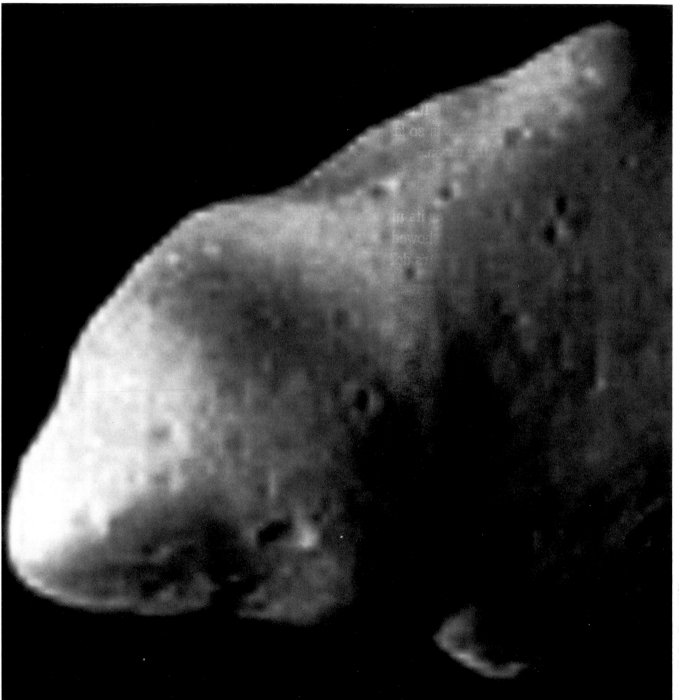

WATCH OUT FOR THOSE ASTEROIDS!

Do you like rocks? How about rocks as big as a house? Or as large as the World Trade Center towers in New York? Or how about a rock almost as large as the *state* of New York? There are such rocks in the solar system and they are called asteroids. They make up the Asteroid Belt—a circle of boulders and rocks that lie just past the orbit of the planet Mars and before that of the gas giant Jupiter.

There are thousands, perhaps hundreds of thousands, of rocks in the Asteroid Belt, ranging in size from miles in diameter to grains of dust. All of them orbit around the Sun, and none of them has an atmosphere.

Rough Rocks

Look around the Asteroid Belt. Asteroids are much different than the rocks in our backyards or other colorful rocks and minerals found on Earth. The darker asteroids are probably made of carbon materials; the brighter asteroids may contain such metals as iron and nickel. How do we know the composition of asteroids? By analyzing chunks of rock from space that strike the Earth, called meteorites. Most of these rocks probably came from the Asteroid Belt (although others may have come from the Moon and Mars).

Let's explore the larger asteroids. You will notice that it is not easy to walk on them. Some are covered with a dusty soil; others are pitted and rough. How much would you weigh on an asteroid? Not much! If you weighed 100 pounds on Earth, you would only weigh 5 pounds on Ceres, the largest asteroid. In order to explore an asteroid, you would have to hitch yourself to the rock like a mountain climber!

Not all asteroids stay in the Asteroid Belt. Sometimes these wandering rocks are affected by the gravitational pull of the other planets and satellites. This can push an asteroid into a weird orbit; or one asteroid can collide with another in the Asteroid Belt, causing them one or both to fly off into strange orbits. They still will orbit the

(Opposite) Gaspra is the only asteroid that we have a picture of in the entire solar system. Notice how it looks like Phobos and Deimos, the moons of Mars.

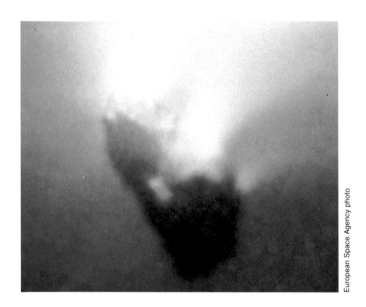

In 1986, the spacecraft *Giotto* visited the famous Halley's Comet, taking this picture of the comet's rock-and-ice nucleus.

Sun, but not between the orbits of Mars and Jupiter. Some asteroids even come close to the Earth as they spin around the Sun.

Hit from Above

These rocks may not be as large as planets and satellites, but they are just as important. If a large (or small) asteroid does stray from the Asteroid Belt, it can eventually cross the path of a planet or

HOW DID WE FIND THE ASTEROIDS?

In the late 1700s, astronomers thought that there might be a planet orbiting between Mars and Jupiter. In 1801, Italian astronomer Giuseppe Piazzi, making a star catalogue, noticed a starlike point. The next night, the point of light had moved, not like a star, but like a planet. Piazzi had discovered the first asteroid, Ceres. It took astronomers a year to confirm his sighting.

Since that time, astronomers have found and catalogued more than 5,200 asteroids, and they believe there are many more. Most of the asteroids are found in the Asteroid Belt; but some others are found in strange orbits. For example, asteroid Icarus often comes close to Earth as it travels around the Sun.

moon of the solar system. If it passes by just at the right moment, there could be a collision.

Asteroid collisions have happened to all the planets and satellites in the solar system. How do we know that it happened on Mercury, Venus, and Mars, and even on the Martian moons, Phobos and Deimos? Because there are huge craters on these solar system bodies, caused by asteroid strikes over the past 4.6 billion years, or since the solar system was born.

Icy Rocks

As we glance around the Asteroid Belt, the rocks look a great deal like other smaller bodies of the solar system called comets. Cousins of the asteroids, comets are made of ice, mud, and dust, and swing in large, regular orbits around the Sun.

We have seen pictures of comets, especially as they move around the Sun. Most comets develop a long tail as the Sun's heat boils off the comet's dust and gases. Some comets regularly return to the inner solar system. Halley's Comet has been a regular visitor every 76 years.

What would it be like to visit a comet? The central part of most comets would be rock and ice. The tail would be dust and gas, and we could easily pass through the tail without even noticing!

HOW DID THE DINOSAURS DIE?

We know that large and small dinosaurs lived on the Earth for millions of years—and suddenly disappeared. Some scientists believe that around 65 million years ago, a shower of asteroids and comets struck the Earth. As the huge chunks of rock and ice fell on our planet, dust and debris flew into the air. The dust was soon carried around the Earth, blocking out some of the Sun's bright light and heat. The temperature changed. The dinosaurs' world grew dark and cold. Over time, because the dinosaurs could not adapt, they became extinct.

Could this theory be true? Some astronomers say yes; others say no. But we know asteroids have struck the Earth in the past: just look at Meteor Crater in Arizona, or the Sudbury crater in Canada.

Chapter Five

THE JUPITER SOLAR SYSTEM?

Jupiter's night side is often filled with auroral displays and large bolts of lightning. The bolts of lightning have much more electrical energy than the most powerful lightning on Earth.

(Opposite) Jupiter is the largest planet in our solar system. It is known for its banded belts and many spots and ovals, which are really storm systems that swiftly travel around the planet.

Let's turn to the planet Jupiter. It is difficult even to imagine a planet the size of Jupiter. As the largest planet in the solar system, Jupiter would fit more than 1,400 Earths inside it! If we ran around the circumference of Jupiter at 6 miles an hour, it would take us 1,935 days to finish our trip; on Earth, such a run would only take 173 days.

This Is Big!

Jupiter takes about 12 years to spin around the Sun, much longer than a year on Earth. In fact, if you were 10 Earth-years old, you would not have even reached your first Jupiter birthday!

But the days would go fast. A Jupiter day lasts only about 10 Earth-hours. If you looked into the sky every hour, you would be surprised at how fast the planet's 16 moons cross it.

What do we see when we look at Jupiter's atmosphere? As we get closer to the planet, we can see colorful gases that are always moving. The planet

is like a huge cloud machine. Its air is mostly made of hydrogen and helium, with other gases giving the clouds their characteristic pastel and deep, rich colors.

These swift-moving clouds are probably responsible for the crackling arcs of lightning that work their way across the clouds. Standing in a lightning-filled area, we can hear the deafening roar of wind, but on Jupiter—unlike Earth—there is no rain!

Ahead of us we can see the Great Red Spot, a giant "storm" that has been taking place on Jupiter for more than 100 years. The bright red spot is

(Top) Jupiter's southern hemisphere contains the Great Red Spot; also included in this picture are the moons Io and Europa.

(Below) The Earth is very small compared to Jupiter. Notice how it is even smaller than the Great Red Spot, one of the major storms in the gas giant's atmosphere!

The rings of Jupiter are very thin and delicate. It looks as if the rings stop, but they are really just in the shadow of the giant planet.

more than three times the diameter of Earth and is surrounded by wispy swirls of clouds in almost every shade of orange, blue, green, and yellow.

The upper cloud layers are the coldest, measuring around −238 degrees Fahrenheit (−150 degrees Celsius). And as we go deeper into the atmosphere, it warms to around 68 degrees Fahrenheit (20 degrees Celsius), close to the average temperature on our own planet. But as we make our way towards the middle of the planet, it becomes darker and warmer, and the air pressure increases to unbearable levels.

Something else is very different about Jupiter: we can never really step on this planet because there *is* no hard surface as on Mercury, Venus, Mars, and Earth! Jupiter is made up of gases—helium, hydrogen, methane, and ammonia. At its center is a small core formed by the gases under enormous pressure. If we could somehow stand on this core, our arms would feel heavier than lead. The gravity would make our legs buckle. We would not be able to take even one step because we would weigh so much. It would feel like walking around with a whale on your back!

As we back away from the gas giant, we also notice that Jupiter has faint rings! We can see three small rings, made up mostly of dust particles. How did the rings form? One guess is that the rings are the remains of a moon that was ripped apart by the gravity of Jupiter and its larger moons

Jupiter's moon Io is one of the most active bodies in our solar system. It is being torn apart by the closeness of Jupiter and the moon Europa.

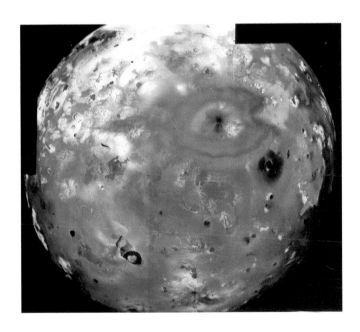

When the *Voyager* spacecraft flew by Jupiter's moon Io, it recorded up to 9 active volcanoes. Can you see the eruption in this photo? (It's right on the horizon.)

Io has light and dark spots, which show the hottest and coolest spots on the tiny moon.

What a Hot Spot!

Whoosh! We had better watch out as we approach this moon; it is one of the most active in the solar system. It is called Io, and it is constantly throwing volcanic material into space. It also is one of the larger moons around the planet Jupiter.

If we traveled to Io, we would notice that all is not quiet. Like a restless sleeper, the moon seems always to be in motion. Why? Because, at one time, it is pulled in the direction of its near neighbor, the moon Europa; at other times, it is being yanked towards its mother planet, Jupiter. In re-

sponse to this movement, the moon has more than 9 volcanoes continually spewing hot sulfur lava and tons of sulfur dioxide gas miles above the surface! In other words, Io is slowly being torn apart.

Let's take a closer look at Io. Nothing can live on it, so we will not see plants, green grass, or oceans. Because of the constantly changing surface, we will not even see craters as on other moons.

We notice that Io is just a little larger than our own Moon and is moving fast around Jupiter. It takes just over a day and a half for it to travel around the giant planet. Because of its volcanic activity, Io also generates electrical power—more energy than from 100 electrical power stations on Earth!

The surface temperature of Io is about -243 degrees Fahrenheit (-153 degrees Celsius) where there are no volcanoes. But if we were able to walk on the surface, we would need a heat-resistant space suit. Why? Because there are so many volcanoes heating up the surrounding atmosphere!

As we look around, we see plumes of volcanic explosions around us and along the horizon. It's a beautiful sight, and not at all like volcanic eruptions on Earth. Since Io does not have much

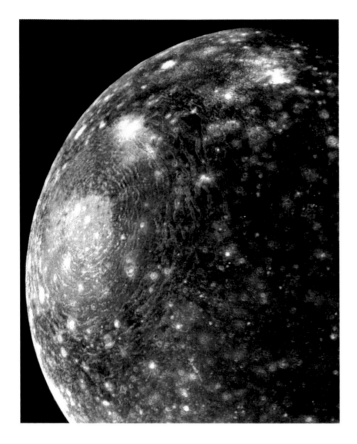

Jupiter's moon Callisto has been quiet for millions of years, unlike many other Jupiter moons. It is marked by hundreds of craters across its surface.

The second-largest moon, Callisto, is the most heavily cratered moon in the solar system. One of the craters, the Valhalla Basin, is a circular area about 373 miles (600 kilometres) across and surrounded by 15 rings! From above, it looks like the ripples that form when you drop a stone into a pond.

Our next stop is the satellite Europa, where your space boots would crack and crunch on the thick ice. Europa is just a little smaller than our own Moon and, unlike most members of the solar system, has very few craters. Instead, it is covered with ice and crisscrossed by dark cracks. As we look around, we wonder how the cracks formed. One guess is that when a space body strikes this moon, it cracks the ice. Mud from underneath the ice then seeps into the cracks— almost like toothpaste being squeezed from a tube!

As we visit the other twelve moons that circle Jupiter, we notice that many are strangely shaped, much like Phobos and Deimos, the moons of Mars. Could they, too, be captured asteroids from the nearby Asteroid Belt? No one really knows. Some of the satellites spin close to Jupiter and others are found in orbits far from the planet, all dancing around the gas giant like a miniature solar system.

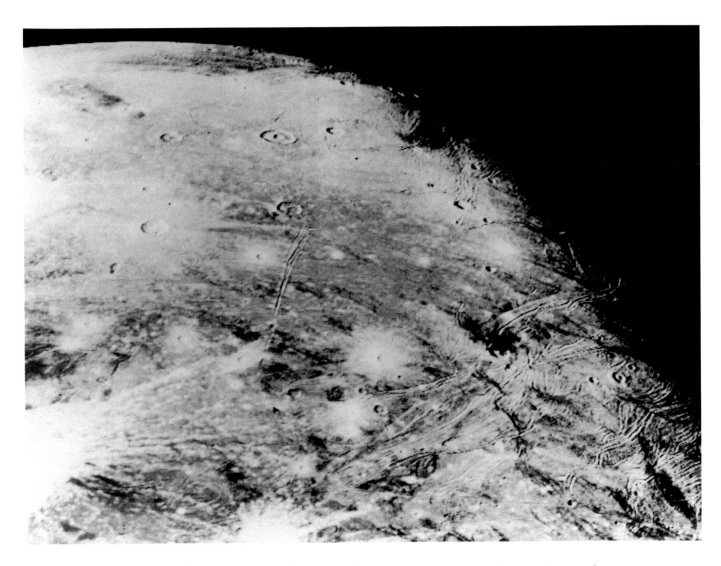

A close-up of Callisto shows cracks and craters that crisscross its half water-ice surface.

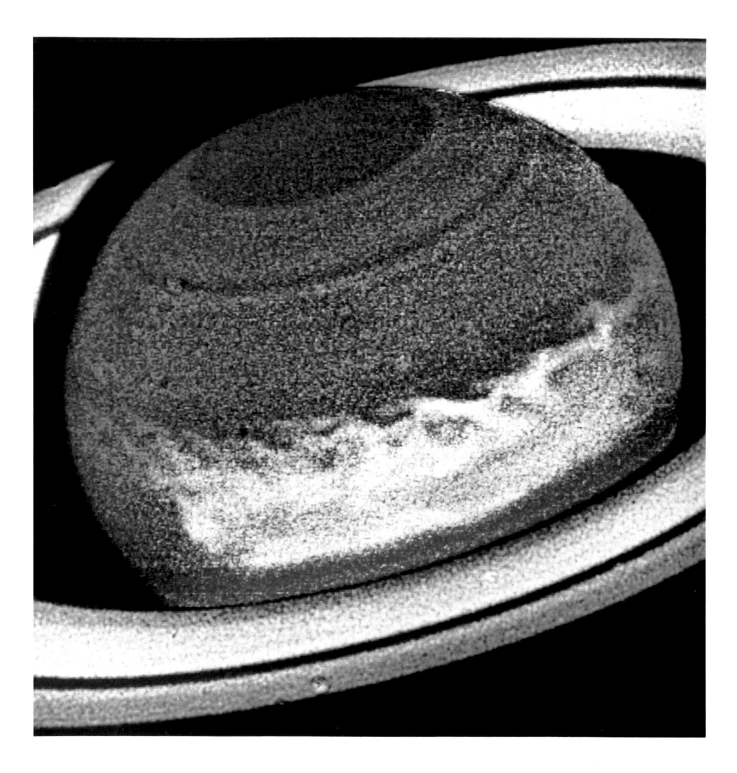

Chapter Six

THE RINGS AND MOONS OF SATURN

The second-largest planet in the solar system is Saturn. We can see it is unique: it has the largest set of rings of all the planets. It also has pastel-colored clouds, and its interior is much like Jupiter's—all gases and no real surface. Because Saturn is mostly gas, it is also very light. If we could find a large enough, water-filled bathtub, the whole planet would easily float!

Saturn is like Jupiter in many ways. But Saturn's winds move more rapidly around the planet, some up to 1,000 miles (1,610 kilometres) per hour. If we had the same wind speeds on Earth, nothing would be left standing!

Saturn's air is mostly hydrogen and helium; other gases give the clouds their characteristic light colors. It has storms that seem to stay in one spot, but they are not as large as Jupiter's Great Red Spot. Giant storms can pop up suddenly on

Saturn was the farthest planet known to the ancients. It is also the only planet whose rings can be seen from Earth with Earth-based telescopes.

(Opposite) In 1990, the clouds of Saturn "burped"! A white storm appeared around the planet—a sudden storm that has since died down.

Biggest Moon

Swing around Saturn and you will immediately notice its moon Titan. This giant moon is larger than our Moon or even the planet Mercury. Its atmosphere is made up mostly of nitrogen just as on Earth. Why can't we breathe on Titan? Because, along with the nitrogen, the atmosphere contains gases which are poisonous to human life.

What do we know about Titan? It orbits Saturn every 16 days and its temperature may go down to −200 degrees Fahrenheit (−129 degrees Celsius). We also know that there is a thick layer of reddish smog covering it. Titan also may have oceans—not like our water oceans, but pools of liquid nitrogen and methane. Would there be winds? Would there be ice? We still don't know enough about Titan to say.

Moons, Moons, and More Moons

Look around. Did you notice the other 23 moons orbiting Saturn? There may be more, but we won't know until more spacecraft are sent to the ringed planet. The other moons of Saturn are not as large as Titan and do not have atmospheres. Some are dark; some are light. And some are "caught" in Saturn's rings.

Many of the moons are lined with grooves or marked by craters. Astronomers believe that

The largest moon around Saturn is Titan. This moon may also be one of the only moons in the solar system with an atmosphere, here seen as haze along the moon's horizon.

many of them are all that's left of a larger moon that split apart in a huge collision.

One moon is particularly striking: it is a small chunk of rock called Mimas. Although it is only 249 miles (400 kilometres) in diameter, Mimas has a large crater that measures one quarter of the moon's diameter. The crater's walls rise higher than Earth's tallest mountain, Mount Everest!

Iapetus is another small satellite. Dark and rocky on one side, it is white and icy on the other. Hyperion is a battered piece of rock that may be a leftover from a larger moon that once circled Saturn. Rhea, the second-largest moon of Saturn, is covered with craters, while Enceladus has some small craters and other areas with no craters at all.

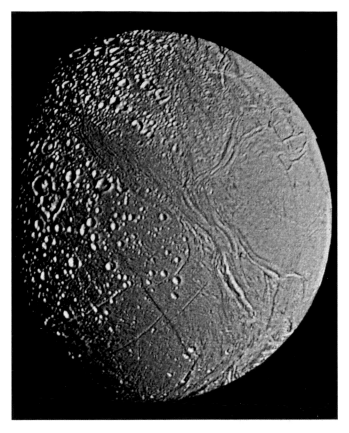

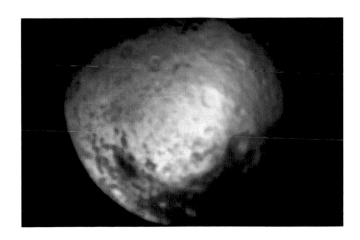

Enceladus may be a very active moon. Notice that there are not as many craters as on the other moons, meaning that there may have been active volcanoes that covered the craters.

(Left) Saturn's moon Iapetus is odd: it is half light and half dark—and no one knows why!

This close-up of Uranus' moon Miranda shows its tall cliffs and grooves.

HOW ARE THEY NAMED?

How are all the moons, craters, and features of the solar system named? They all are named after famous people and characters from well-known books or ancient legends. For example, the craters and features on Venus are all named after women. Some are women from mythology, such as Aphrodite. Uranus' satellites are named after characters from William Shakespeare and Alexander Pope, such as Miranda from Shakespeare's *The Tempest*.

Asteroids and comets are named differently. In most cases, the asteroids are named at the request of the discoverer. (One scientist named an asteroid Spock, after his cat!) Comets are named after their discoverer. Many times the comets get two names, because they were discovered by two scientists working together.

Titania is Uranus' largest satellite. It has several large cliffs and faults which may have been caused as the icy moon expanded and contracted.

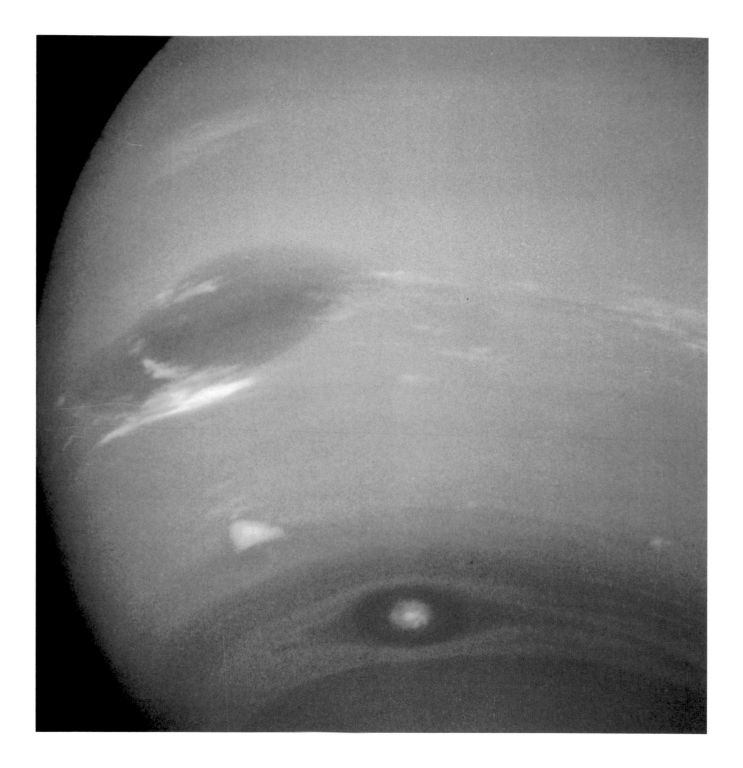

Chapter Eight

IT'S EVEN COLDER ON NEPTUNE

Let's turn to the pale, bluish green planet Neptune. It has broad bands in different shades of blue and bright polar regions. We would definitely notice how dark it is on Neptune: the sunlight is about 900 times dimmer than on Earth. If we look back at the Sun from the planet, it would look like a flashlight beam seen from 300 yards (274 metres) away!

If we were to wait for a year to go by on Neptune, we would be there for 168 years. While there, we would have to keep our space suits on because the atmosphere is filled with hydrogen, helium, methane, and ammonia. Keep the suit's heater on, too; the temperatures are about −353 degrees Fahrenheit (−178 degrees Celsius).

Probably the most obvious features we see as we look at Neptune are spots in its clouds. These are great storms. The Great Dark Spot, the Small

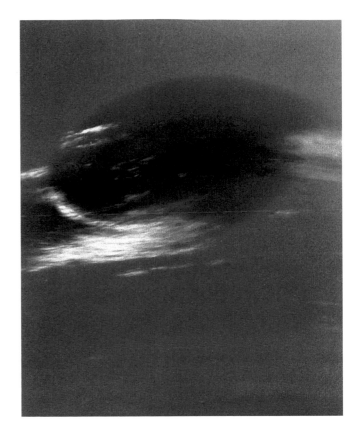

The Great Dark Spot is a storm in the top clouds of Neptune, similar to the Great Red Spot on the planet Jupiter.

(Opposite) **Neptune may have close to the same bluish color as Uranus, but it is filled with cloud features.**

PLANET CHART

Planet	Diameter (miles/kilometres)	Mean Distance from the Sun (million miles/kilometres)	Length of the Day	Around the Sun
Mercury	3,030/4,880	36.0/57.9	59 days	88 days
Venus	7,521/12,104	67.2/108.1	243 days	225 days
Earth	7,926/12,756	92.9/149.5	23 hr, 56 min	365 days
Mars	4,217/6,787	141.5/227.9	24 hr, 37 min	687 days
Jupiter	88,730/142,800	483.3/778.3	9 hr, 50 min	11.9 years
Saturn	74,900/120,540	886.2/1,427.0	10 hr, 39 min	29.5 years
Uranus	31,763/51,118	1,783.1/2,869.6	17 hr, 54 min	84 years
Neptune	30,775/49,528	2,794.0/4,496.6	19 hr, 12 min	165 years
Pluto	1,430/2,300	3,666.0/5,900.0	6 days, 9 hr	248 years

GLOSSARY

asteroids: Small rocky objects in the solar system, ranging in size from around 600 miles (1,000 kilometres) across to dust particles.

asteroid belt: The region of the solar system where we find the majority of asteroids, in a belt between the orbits of Mars and Jupiter.

carbon dioxide: A colorless, odorless gas; it is common in the atmospheres of some planets, including Venus and Mars.

channels: Long, winding cuts in the surface of a planet or moon; for example, river channels on Earth.

comets: Smaller members of the solar system, made of ice, rock, and gas, that orbit the Sun.

craters: Large and small holes on a planet or satellite, usually caused by the impact of asteroids, comets, or meteorites.

crescent: The phase of a planet (or our Moon) where we can only see a slice (or crescent) of the body as it is lit by the Sun.

double-planet system: Where a planet and its moon closely orbit each other, and where the moon is relatively large compared to the planet. Pluto and Charon, and Moon and Earth are often considered double-planet systems.

dust devil: A tall column of dust that moves swiftly across a surface, such as on Mars during a dust storm.

"evening" star: The planet Mercury or Venus when either planet is seen in the evening sky from Earth.

Galilean satellites: The four largest moons around Jupiter: Io, Ganymede, Europa, and Callisto.

gravity: The force that pulls two objects together; the larger the planet or satellite, the greater the gravitational pull.

greenhouse effect: In astronomy, usually referred to in association with Venus, where the clouds are so thick that the Sun's radiation is not allowed to bounce off the planet and into space. The term comes from the glass-constructed greenhouse, which traps and holds in heat.

ice caps: Patches of ice seen at the poles on several of the planets and satellites of the solar system. They usually grow in size during the planet's winter and shrink in the summer.

meteor: A piece of rock or dust that enters the Earth's atmosphere and burns up. We see it as a brief, bright trail in the nighttime sky.

meteorite: A piece of meteor large enough to have reached the Earth's surface. The largest found so far weighs 132,000 pounds, and is located in southwest Africa.

moon (*see also* **satellite**)**:** A natural celestial body having a regular orbit around another larger body, such as a planet.

"morning" star: The planet Mercury or Venus when either planet is seen in the morning sky from Earth.

orbit: The path followed by a body moving around another body. For example, our Moon is in orbit around the Earth, and the Earth is in orbit around the Sun.

ring: An orbiting stream of large and small particles around a planet. Four planets in our solar system have rings: Jupiter, Saturn, Uranus, and Neptune. We do not yet know if Pluto has rings.

satellite (*see also* **moon**)**:** A natural or artificial body having a regular orbit around a larger body.

solar system: The collection of planets, moons, asteroids, comets, and other chunks of rock that circle our Sun.

volcano: A cone-shaped mountain or other opening in a planet's crust that spews out steam, gas, ash, and often hot lava. The Earth and Io are two members of the solar system with active lava volcanoes.

INDEX